To my dear friend Jean ~ :)
God's blessings on you!

D1576121

Priscilla Mitchell :)

SEEDS

God's Awesome Computers

Priscilla Mitchell

WINEPRESS WP PUBLISHING

© 1998 by Priscilla Mitchell. All rights reserved

Printed in Korea

Packaged by WinePress Publishing, PO Box 1406, Mukilteo, WA 98275. The views expressed or implied in this work do not necessarily reflect those of WinePress Publishing. Ultimate design, content, and editorial accuracy of this work is the responsibility of the author.

No part of this publication may be reproduced, stored in a retrieval system, or transmitted in any way by any means—electronic, mechanical, photocopy, recording, or otherwise—without the prior permission of the copyright holder except as provided by USA copyright law.

Unless otherwise noted all scripture quotations are taken from the New King James Version, Copyright © 1979, 1980, 1982 by Thomas Nelson, Inc., Publishers. Used by permission.

ISBN 1-57921-128-3
Library of Congress Catalog Card Number: 98-60806

Dedicated to God's glory;

to my parents, Byron C. and Anita P. Nelson, who
were faithful to His Word; and to my husband, Richard,
who consistently supported this work.

～ Acknowledgments ～

Much gratitude goes to precious friends who faithfully inquired about the progress of the poems, encouraged the author with enthusiastic words, and prayed for success in the work.

Early help came from Betty Chandler and sisters Betty Taylor and June Solberg with input on meter, and critiquing by published friends.

Thanks also goes to Fred Fortney, Nan Heim, and Paul Heser for using their special skills when needed.

Dr. Don Carson, professor at Trinity International University and former neighbor, reviewed my poems in an earlier form and sent me back to the drawing board. I thank him for his time and expertise. As with surgery, it's painful but necessary.

Jeanette Phillips' amazing phone call from Florida brought assurance to me that God was watching over all. I thank her for her help, patience, and love. Perhaps one day we'll meet face to face.

Margaret Wiggins' note, "*You Can Do It,* Athena Dean," and son Mark and daughter-in-law Sopheap's message with a 1-800 number a few days later, set me on the WinePress Publishing road. I thank them for keeping my book goal in mind.

My son Dan was a great encourager in the work of self-publishing. "We can do it!" he said, and offered his equipment and office for my use. He was there, patiently helping this novice with her frequent computer glitches.

Also supporting with computer expertise, but especially intercessory prayer, was dear friend Irene Wyckoff.

There are not adequate words to thank the eight illustrators for their labor of love. Most of them had busy schedules, but agreed to use their gifts for this work: Mary Campbell, Marty Engel, Amy Haley, David Mitchell, Allison Quane, John Scott, Ron VanDerLip, and Terry Rawlings—whose patience over the long haul has been such a blessing.

Finally, my gratitude to Richard, my husband, for loving and encouraging me through it all.

∼ Contents ∼

A Note from the Author

Have you ever studied the columbine? In 1987, as I examined this beautifully detailed flower in our garden, I marveled, "How do those little parts know which way to go? Surely God has programmed the whole process! Children are impressed with computers. They ought to realize God had His computers a long time ago."

I went in the living room, to my favorite quiet place, and began to write down those inspirations. The first two lines rhymed, and so thus began this work about seeds.

The more I ponder each growing thing, the more I'm in awe of the One Who could do all this from such tiny creations. I hope you feel the same way. My heart's desire is to help children realize what a marvelous Creator and Savior they have.

Seeds:

God's Awesome
Computers

Have you held a seed in the palm of your hand?
Have you stopped to think
　　　　　　　how these small seeds were planned?

The land holds so many!

The sea has them too.

They follow God's orders and know what to do.

Much like a computer game programmed to play,
each seed waits the time for its growing display.

An egg with its seed also carries inside
the tightly packed plans of its Heavenly Guide.

Now think of the programming God must have done
to make each new seed and decide, one by one,
just how it should grow, how big it should get,
and where it should live, in a place dry or wet.

A flower perhaps?

A tree mighty tall?

Or maybe it won't form a plant after all,

but rather . . .

a fish—with a cover of scales

or something with skin,

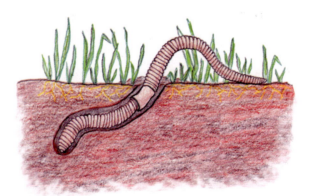

or a shell hard as nails.

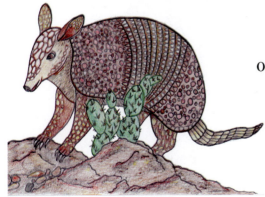

Yes, think of the covers God makes from the seed,
the kind that He plans for each life's special need.

Most birds have fine feathers to help in their flight,

an elephant's thick hide can suit him just right,

sharp quills for the porcupine,

bark for the tree,

and plating so thin for the horse of the sea,

warm wool for the sheep,

while the bears have their fur,

soft hair for the kittens

that mew, hiss, and purr.

Oh, think of the sounds that God plans from the seed,
to send out a message that others will heed:

a buzz for the bee

and a bark for the dog,

a chirp for the cricket,

a croak for the frog,

a moo for the cow,

and a baa
for the sheep,

a crow when the rooster
awakens from sleep.

The trees are all filled with such beautiful song
from bright colored birds singing all the day long.

Let's think of the colors God makes from the seed,
to brighten His world, adding beauty indeed!

The tan and the white of a shy, gentle fawn,
the green of the grass as it carpets the lawn,

the orange of sunfish that swim in the brook,

the blue of the peacock that struts when we look.

Some flowering plants dressed in glorious clothes

have wonderful fragrance that pleases the nose.

Just think of the smells that God plans from the seed
to add to the pleasure that He has decreed.

The lovely white lily,

the lilac, and rose,

the blossoming clover,
and others like those

have perfume so special it's hard to compare.
Each seed holds inside its own recipe rare.

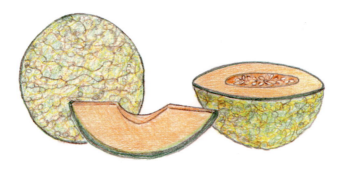

The smell of sweet melon or corn popped by heat

adds much to the taste that makes these such a treat.

Now think of the tastes God provides from the seed,
to bring great enjoyment to all that may feed

on apples and strawberries,

grapes from the vine,

on honey and peanuts with flavors so fine.

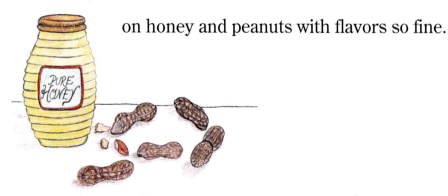

Some foods are made salty, some foods are made sweet, and some are made tart or quite spicy to eat.

The horses like hay,

there is nectar for bees,

and berries for some birds with nests in the trees.

Yes, think of the houses designed from the seed,

and methods for travel with fast or slow speed,

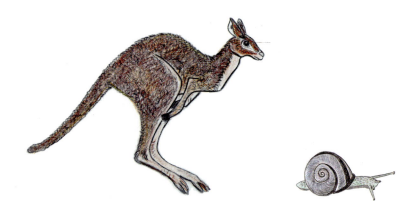

and times for the flowers to open each day,

and number of offspring with their kind of play.

The food they enjoy and the way that they feed
are guided along by each marvelous seed.

Can you add to this book your own list of ways
a seed shows God's planning and how it obeys?

We live in a world filled with wonders untold.
God's seeds follow orders, then wonders unfold.

Each seed is a picture of purpose and plans.
These awesome computers are greater than man's,
for looking around we're delighted to find
God's awesome computers make more of their kind.

Since God has made plans for each marvelous seed,
just think how He knows about *your* every need.

The
Delightful
Dolphin

Have you ever seen the dolphin

    as he frolics in the sea,

        as he leaps and dives through blue-green water

        deep as deep can be?

        He's graceful, sleek, and streamlined too.

      His tail—prepared for power—

   has muscles pulling up and down

through every fun-filled hour.

He swims with playful cousins,
        and they race along so fast,

they look as if they're laughing
        at some big boats they have passed.

These ocean ships sail smoothly
        as they're guided on their way

by what we now call *sonar*.
        It can lead them night or day.

But God first made the dolphins

with this marvelous design.

Why, they can speed through unknown ways

and never need a sign.

The sounds they send out bounce back

from each object far or near.

This echo guides or warns them

through their ever-listening ear.

The dolphins show us what it means
to always help another.

They rescue any wounded friend and save a tired brother,

for God has put within the seed
that made this creature smart,

a gentle, kind, and friendly way. He made a caring heart.

Now enemies might swim about.

The dolphins then will know
they must watch out and guard the weak.

Again their help they show.

When time arrives for baby calf to join its family,
the mother has protection from big sharks and injury.
And did you know, at time of birth,
the baby needs his air?
Another dolphin comes along, if help is needed there,

and lifts the baby—born tail first—
above the ocean waves.
By doing so—you see God's plan?—
the baby's life she saves,

for baby has no gills like fish.

It must breathe air like man.

But how will he take in this air?

Someone arranged a plan!

The air comes through the blowhole

God prepared upon his head,

but when the water splashes near,

it closes up instead.

Since God designed this baby calf
to be in mammal form,

it nurses near the mother's tail,
and drinks her milk so warm.

When mother calls, he knows her voice
and answers with his own.

They talk with grunts and whistles.
He can hear her softest tone.

When he is almost two years old
he leaves his mother's side

to join the older dolphins
as they swim the ocean wide.

We find delight in watching them
                at many water shows.
How can they learn to do those tricks?
                Their Great Creator knows!

His ways are truly awesome!
                His designs show special care.
The life-seed of the dolphin
                has the plans already there.
Since He has planned for mammals
                that can frolic in the sea,
just think of all the plans He
                has in store for you and me.

Have You Thought About . . .

How generous God is? How we put one apple seed in the ground and get a tree full of apples? And each apple has how many seeds? God is great at math! He multiplies all the time. His giving is so much better than ours. Think of some other examples.

How God enjoyed making His whole creation? He gave us corn that explodes in the heat, tall birds that do silly-looking dances, monkeys that make faces, and more. He made animals to show us special qualities, like the kindness of a dolphin, the faithfulness of a dog, the stubbornness of the mule, and the laziness of the sloth.

How God makes great food packages from seeds? He wraps His corn in overlapping green leaves, and inside we find His yellow kernels (seeds) in such neat rows. He gives His oranges thick skins to hold all that juice in small, pointed containers. Then He makes double walls to give us nice segments (packages) to hold. Think of the pods for peas and beans. At your supermarket, look for more examples.

How God forms bones out of soft cells? Some cells will make teeth or nails, horns or hoofs. How can this be? How do the teeth and bones know when to stop growing? What would happen if they didn't?

How God made us special—in His image? He made us creative. We can build a brick house, an igloo, a skyscraper, and more. The beaver makes his dam, the bird makes her special kind of nest, the spider makes its web, but God lets us think and design what we'd like. Now we can even make computers, but remember, God programmed seeds, His "awesome computers," a long time ago.

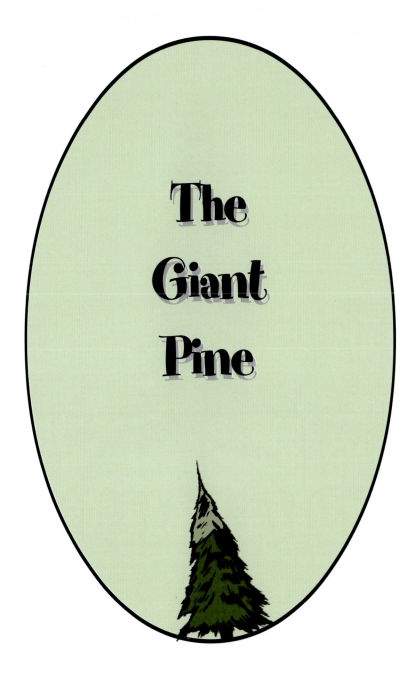

The
Giant
Pine

A dark green, giant pine tree

      has been living in our yard.

      It stands there tall and straight

            just like a military guard.

Below its spreading branches

do you know what can be found?

Some rough, brown, woody pine cones

lying scattered all around.

These cones contain the small seeds

of this beautiful big tree.

How different they appear now

from the pine trees they will be!

But how do little pine seeds
find their way out of the cone?

There's one way frisky squirrels make,
who won't leave them alone.

They chew and tear the cones apart
to have a tasty treat.

The seeds they have discovered here
are something good to eat.

But other pretty pine cones
that cling tightly to the tree

will open up, the wind will sway,
the seeds will then fall free.

Now some of those will drop
into a good place on the ground,

and by the power in those seeds
new shoots will soon be found.

They'll grow into great pine trees

of the Family Evergreen,

with needles long and shiny

but as soft as you have seen,

or maybe they'll be short and prickly.

This will all depend

on what instructions there will be,

what orders they will send,

for each small seed has been designed—
directions kept inside—
with wisdom, planned and programmed
by a caring Heavenly Guide.

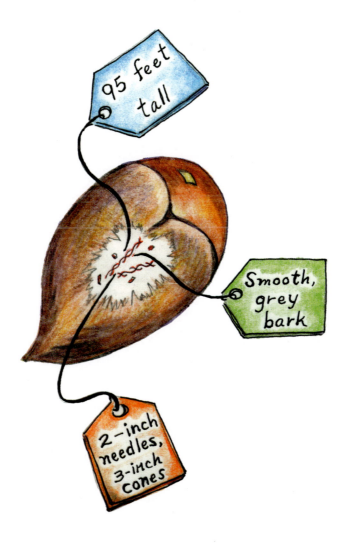

It tells the tree, from year to year,
how tall the trunk should grow,
what kind of branches it should make,
what bark or leaves to show.

What does it mean to be a part
of Family Evergreen?

With red or yellow dropping leaves
it never will be seen,

and late in winter, when most trees
look dead, quite bleak and bare,

the scent of green pine needles
puts a perfume in the air.

To all of God's own people
a sweet message it can send:

"My green is a reminder
that your life will never end.

My branches, too, can do their part.
They hold the Christmas lights,

which help you celebrate each year
the Holy Night of nights."

Now who should get the credit for the seed so very small
that made the pine tree by our house
which grew so very tall?
The Lord of earth and heaven,
Who created all things good,
Who wants us to enjoy His work
and world just as we should.

The wonders of His trees include
much beauty and nice shade,
good food, and every sort of wood,
and much more He has made.
Well, this is cause for giving thanks,
from hearts warmed up with love,
for gifts so great and generous from Father up above.

The Hummingbird

Outside the cabin window there is something fun to see.

She's sipping sweet red liquid

from a feeder in the tree.

She flies about so quickly

as she darts from place to place,

you must pay close attention just to see her tiny face.

Her wings are always fluttering
in such a speedy way

you hardly see them moving
when you watch her in her play.

I wonder how this creature,
not much bigger than my thumb,

can fan her wings
with such a beat—
it sounds quite like a hum.

Was she designed and programmed
by a God Who is all wise,

Who wants us to enjoy this
pretty creature as she flies?

Before the early makers
of our airplanes had a plan,
this little one called "hummingbird"
was seen by many a man.

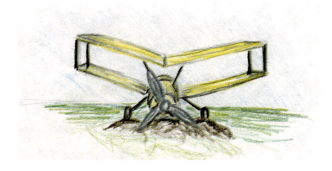

We see her flying forward, upward,
downward—and what's more,
she also can fly backward from the hanging feeder door.

Her Maker has her hover
at a flower where she sips.
A helicopter cannot match
this birdie's expert dips.
Her long thin bill and hollow tongue
were perfectly prepared
to do their busy work all day.
They show a God Who cared.

When time for nesting comes along,
a tiny bed is made,

and mother hummingbird just sits
until two eggs are laid.

They're white and small (of course they are,
she's such a little thing)

and though she's proud, you'll hear no song.
She wasn't made to sing.

Each tiny egg that holds a seed will hatch,
and there we'll find

God's neatly packed and programmed seed
made one more of its kind.

No way this bird will grow by chance

or accident, I'd say.

Its life shows careful planning

for each step along the way.

I think I want to

praise this God

Who must be marvelous.

If He can plan

for little birds,

He must have plans for us.

The Shy
Sea Horse

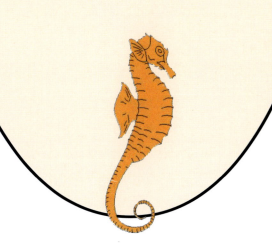

The sea horse isn't a horse at all.

It's rather a fish that is somewhat tall,

with a crown of bumps on a horselike head,

with fins that move him straight ahead

while standing up (or so it looks),

with twisting tail that loops and hooks.

How can there be a fish so strange?
How can he make his color change
when frightened by some danger near?

His Maker planned him, it is clear.
Why, he can match a shell or stone
when wishing to be left alone.

Now most fish have some scales or skin,
but his hard shell, though sort of thin,

is full of spines, a cover strong—
his armor—as he moves along.

The sea horse plays

with his own kind.

In tug-of-war

their tails will wind.

He'll somersault or climb a weed
to watch for food on which he'll feed.

His eyes can look

both ways at once

to see small shrimp,

for which he hunts.

There's something you may find quite odd

about this fish designed by God.

The mother lays her eggs inside

the father's pouch where they will hide.

Then off she'll go, her job all done,

but father's work has just begun.

Inside his pouch the eggs will grow.
His body soon will clearly show

the time is right! God programmed it.
A herd of fish, all formed and fit,

will tumble out to swim about.
A wondrous thing, there is no doubt!

No lessons at a nearby pool,
no classes at the fishes' school!

They simply do what God had planned
and follow His divine command.

All creatures of the sea are His.
Oh, praise the Mighty God, Who *is*!

Since He designed the shy sea horse,
does He have plans for us? Of course!

Scripture to Memorize

Psalm 50:10–11: For every beast of the forest is Mine, and the cattle on a thousand hills. I know all the birds of the mountains, and the wild beasts of the field are Mine.

Revelation 4:11: You are worthy, O Lord, to receive glory and honor and power; for You created all things, and by Your will they exist and were created.

Jeremiah 23:24: "Can anyone hide himself in secret places, so I shall not see him?" says the LORD; "Do I not fill heaven and earth?" says the LORD.

Psalm 139:13–14: For You have formed my inward parts; You have covered me in my mother's womb. I will praise You, for I am fearfully and wonderfully made; marvelous are Your works, and that my soul knows very well.

Psalm 119:11: Your word have I hidden in my heart, that I might not sin against You.

John 14:6: Jesus said to him, "I am the way, the truth, and the life. No one comes to the Father except through Me."

John 15:12: This is My commandment, that you love one another as I have loved you.

John 14:3: And if I go and prepare a place for you, I will come again and receive you to Myself; that where I am, there you may be also.

Revelation 2:10*b:* Be faithful until death, and I will give you the crown of life.

The Funny Monkey

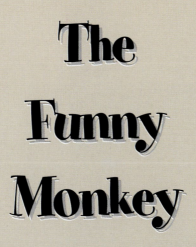

Hello, funny monkey,
up there in the tree!

Your swings and your jumps
look so easy and free.

You hang by your long tail
or climb with your toes,

and how do you do this?
Your wise Maker knows!

He placed in the seed that makes
creatures like you

His marvelous program
that guides what you'll do.

He made you an acrobat,

agile and bright.

Your tricks and quick moves

at the zoo are a sight!

You clean up each other

and care for your young.

You make silly faces and

show off your tongue.

You romp or just sit

and examine the others.

Your babies all seem

to cling to their mothers.

Your life in the jungle,
  on branches so high,

    gives you a good lookout
      from where you can spy

      to bark out a warning,
        if trouble is near.

          How peaceful the living
            when there is no fear!

You stay with your family—
some call it a *troop*—

and the strongest of you
is in charge of the group.

You surely are different from our friendly cat.

She, too, has a long tail, but can't swing like that!

God programmed your tail to have bare skin below,

with ridges and muscles to help as you go

in search of some sweet fruit that hangs on a tree.

Your strong tail holds on so your arms can be free.

We see, funny monkey, since God designed you,
your life-seed, so well planned, knew just what to do

to make a smart mammal who, all through the day,
can "monkey around" in his own playful way.

You'll never be changing, you always will be
that cute, funny monkey we all like to see.

Hurrah for the fun that you bring to the zoo,
but the real praise belongs to the One Who made you!

Things to Do

(Name, draw, copy, cut, paste—and more)

1. Draw seven different kinds of seeds.
2. Name or find pictures of five hard-shelled animals.
3. Make a list of the homes animals make for themselves.
4. Learn three bird songs so you can tell them apart. Your library should have a CD of bird songs.
5. Make a booklet of your own like the *Seeds* poem and list more ways a seed shows God's planning.
6. On a trip to the library, look up these things (a librarian can help you):
   a. What are the biggest seeds?
   b. What are the smallest seeds?
   c. What kind of nest does a bowerbird make?
   d. What do seeds look like that make some under-water plants?
   e. What is unusual about a banyan tree?
7. Try to make a bird's nest. How about an oriole's nest? (God certainly made some clever birds!)
8. Copy the designs of the creations below. God is a great designer! His world is full of wonderful variety.
   a. Leaves: maple, palm, and gingko
   b. Beaks: pelican, toucan, and kiwi
   c. Feet: elephant, flamingo, and frog
   d. Coats: zebra, leopard, and giraffe
   e. Wings: bird, bee, and bat
   f. Fruits: pineapple, coconut, and star fruit
   g. Sea shells: zigzag scallop, angel wings, and coquina clam
   h. Bark: oak, pine, and birch
   i. Nests: oriole, ostrich, and osprey

All these things God made through His awesome seeds!

The
Honey
Bee

Now one of the marvels on earth you will see

is the buzzing and hard-working, small honey bee.

Well, you may be thinking about her sharp sting,

but the way she grows up is the wonderful thing.

At first, you will find she's a tiny white dot
that was laid in a wax cell at just the right spot.

This tiny white egg with
her life-seed inside

was carefully planned by
her Heavenly Guide.

She grows and turns into
a chubby white worm,

then under a cover of
wax she will squirm

to wrap herself up in a silk blanket tight.
Her *miracles* happen, but out of our sight.

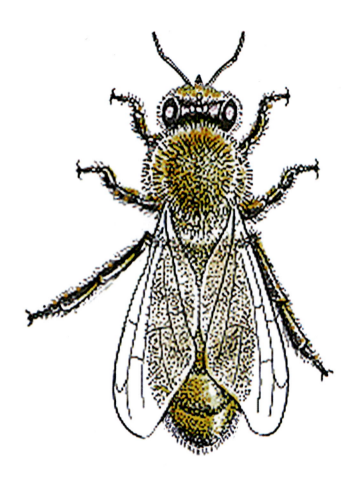

Oh, wonder of wonders! Examine and see!
From out of her cell crawls our small honey bee!

Now, how did she grow to be yellow and black?
And where did she get those four wings on her back?

And the six hairy legs? And that long tubelike nose?
What kind of Designer could plan each of those?

By making her seed to be programmed within,
the parts of her body knew when to begin.

Well, that's just the start of our young honey bee.
  She'll be a hard worker since she is now free.

She'll clean up, then fix up, and build each new cell.
  It must have six sides to it. How can she tell?

But where does she find all the wax she will need?
  It comes from her body—a wonder indeed!

Then working with others in her colony,
  they'll act as a team, like a good family,

by feeding the queen—

and each slow father drone—

and feeding the babies

before they are grown,

by fanning the bee hive to keep it just right,

and finding sweet nectar before it is night,

by bringing in pollen

from beautiful flowers,

then storing this food for

the long winter hours.

And now if our bee finds some food just by chance,
she tells all the others by doing a dance.

She moves in a circle to show if it's near,
but figure-eight dancing will make it quite clear

it's farther away—and then off they will go.
With the dance, the light, and the smell, they will know.

They'll bring home sweet nectar

to put in the cells,

and this becomes *honey*

in each of these wells.

Now, who could design

such a wonderful plan,

with something for insects,

for bears, and for man?

But the sting, you are thinking, is there! Don't forget!

That is her way to say, "I am *upset!*"

Well, who could prepare such a small honey bee
to do all these things? Only God! It is *He!*

Our God, Who can plan what this insect should do,
must also have plans for a person like *you!*

About the Illustrators

Mary Campbell: "Seeds: God's Awesome Computers"
Mary, born in 1964, loves cooking, artwork, and animals. She is a teacher, mother of two, and the only daughter of the author.

Ron VanDerLip: "The Delightful Dolphin"
Ron, born in 1965, is a carpenter who loves the arts. He is a father of two children and enjoys spending time with them.

Marty Engel: "The Giant Pine"
Marty, born in 1934, spent thirty-four years in food engineering research. He's a father of five, grandpa to fourteen, and loves classical music and gardening.

Amy Haley: "The Hummingbird"
Amy, born in 1978, plans to teach when she finishes her college education. Writing, singing, and playing the piano are her special gifts.

Allison Quane: "The Shy Sea Horse"
Allison, born in 1984, homeschooled and our youngest artist, says, "I love the Lord and want to use the talents He has given me to serve Him."

Terry Rawlings: "The Funny Monkey"
Terry, born in 1934, retired as a community college automotive instructor, now serves the Lord in a pastoral ministry. He is father of three, grandpa to five, and faithfully worked as art coordinator for the illustrators.

David Mitchell: "The Honey Bee"
Dave, born in 1956, does counseling, court supervision and conciliation. He is the "first born seed" of the author and writes, "I love my great Creator and am awed by my Savior's creation."

John Scott: "Marvelous Me" and "Wonder Now"
John, born in 1927, works at a cancer treatment hospital and is father of three, grandfather of four. He enjoys the arts and bicycling.

Marvelous Me

When God planned to make a young child like me,

He used a small seed that the eye cannot see.

He added an egg, very tiny indeed,

and made a warm place where I'd have all I need.

This seed in its egg,

with directions so clear,

will follow God's programming year after year.

The one tiny cell becomes two,

and then four.

I'm starting to grow.

The cells multiply more.

Inside of my mother great wonders occur,

but you see a change on the outside of her!

God put in these cells tiny guides we call *genes*,

not clothes like you wear, but His marvelous means

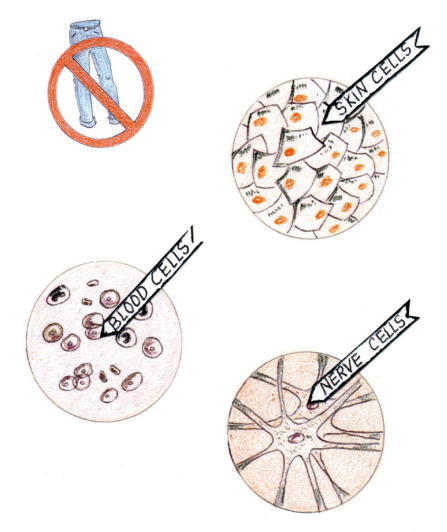

to show every cell special work it should do

to form a small person like me—or like you.

Each cell knows exactly
the place it should go,

the time it should start,
and the way it should grow;

which cells should develop
the heart and the head,

the brain, and the nerves,
and the lifeblood red,

two arms and two legs
 from the backbone strong.

  My marvelous body
   is now growing long!

    One mouth with two lips,
     one nose and two ears,

      can clearly be seen now.
       Say! This baby hears!

The cells that were programmed to make up my eyes

obey their instructions. Our God is so wise.

He gives me these cameras that I may see

the world and the good things that He has for me.

Well, I must have muscles

and these small cells, too,

are busily working.

They know what to do.

For now I am moving,

and even can swim.

God's plan is so wonderful.

I will praise Him!

And now you can count all my fingers and toes,
for God in His planning knew I would need those.

On each little finger He makes my own print
to tell me so clearly (it's not just a hint)

to Him I'm important—His own precious one.
He knows all my days here since they have begun.

My body is formed now,
but that's not the end.

I'll grow and get stronger,
I'll stretch and I'll bend,

until that great day—
when the timing is right—

I come from my mother
out into the light.

And now it's important to see as I grow

that God has a purpose He wants me to know.

I'm not a cold robot. He gives me a choice:

to go my own way, or to heed His kind voice.

His words are so gentle: "Come, walk close to Me,
then with Me in heaven forever you'll be.

Remember, I planned you and know what is good.
Just listen to Me and you'll do what you should,

For My Word will be like the good seed in you
that blossoms and bears fruit in all that you do.

My Spirit will fill you with joy, love, and power,
to give you the help you need through every hour.

I love you, I want you as My chosen one.
To save you from sin, I gave Jesus, My Son,

Who died on the cross and Who now lives above.
Believe in Him, trust Him, and walk in His love!

He's coming back soon in the clouds of the sky,
so you may help celebrate with Him—on high."

O yes, I will follow!

I'll trust Him today!

With His help and guidance,

His Word I'll obey.

I'm glad that He loves me.

My name's written down,

and one day—He promised—

He'll give me a crown!

I wonder now, is there a chance
that all this came by happenstance?

Could eyes be made
by accident,

or ears then hear
a message sent

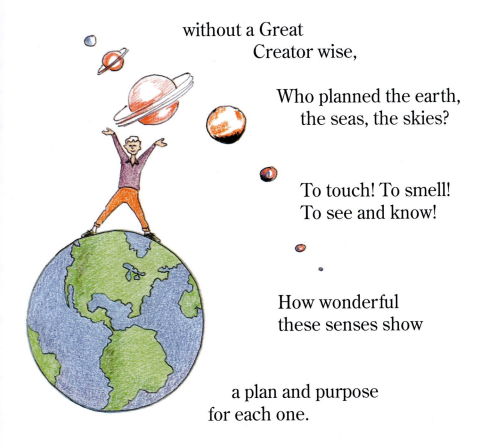

without a Great
Creator wise,

Who planned the earth,
the seas, the skies?

To touch! To smell!
To see and know!

How wonderful
these senses show

a plan and purpose
for each one.

I look and think what
God has done.

I ponder this, and now I find
a world of wonders He designed.

To order additional copies of

Seeds: God's Awesome Computers

send $11.99 plus $4.95 shipping and handling to

Books, Etc.
PO Box 1406
Mukilteo, WA 98275

or have your credit card ready and call

(800) 917-BOOK